D0914271

THE CHEMICAL DREAM
OF THE RENAISSANCE

Churchill College Overseas Fellowship Lectures

Number One
Paradise Lost: a deliberate epic
ERNEST SIRLUCK

Number Two
Why literary criticism is not an exact science
HARRY LEVIN

CHURCHILL COLLEGE
OVERSEAS FELLOWSHIP LECTURE
NUMBER THREE

The Chemical Dream
of the Renaissance

by ALLEN G. DEBUS

W. HEFFER & SONS LTD
Cambridge, England

SBN 85270 000 8

Printed in Great Britain by R. I. Severs Ltd, Cambridge

Allen G. Debus is Associate Professor of the History of Science in the University of Chicago. While Overseas Fellow at Churchill College, Cambridge, in 1967 he gave this lecture in the Wolfson Hall on 2 June

We live today in the midst of overwhelmingly rapid change in all areas of science and associated with this change we find traditional systems of education everywhere subjected to stress on all levels. With this in mind it seems not out of place to turn at this time to the period of the birth of modern science in the sixteenth and seventeenth centuries and to a few of the problems of the educational crisis which arose at that time. Then too there was considerable concern as to how scholars might best study nature and what part the universities should play in educating man about the universe around him.

It is not my purpose to give a history of education in this period and I certainly do not labour under any delusion that a study of the special problems of three and a half centuries ago will solve those of today. Rather, noting the complexities of the present situation, I would call your attention to the fact that the situation was not simple then either. Many histories of science view that age as did Galileo, a dialogue between Simplicio and Salviati—the Ancients and the Moderns. The period is seen as one long triumphal march—from Copernicus to Tycho to Kepler and Galileo and then on to Newton. In reality, however, the period was surely as complex as ours, and alternative solutions to current problems were strongly defended by their proponents. One group certainly, the chemists and the followers of the Swiss reformer, Paracelsus, was convinced that it was not a new study of motion and astronomy which would bring about the sought after reformation of learning but rather chemistry which was thought to be the true key to Nature that would unlock the secrets of heaven and earth. And—they added—if they were right, a revolution in the teaching of the sciences must accomplish their reforms. It is to this group and their plans for educational reform that I would turn.

7

We now know that all of the universities of the sixteenth century were not as reactionary as they are sometimes described. John Herman Randall, Jr., in his 'Development of Scientific Method in the School of Padua' links the fourteenth century critical re-evaluation of Aristotelian philosophy at Oxford and Paris with the northern Italian schools in the fifteenth and sixteenth centuries.[1] Thus a bridge is formed with major figures of the seventeenth century—men such as Harvey, Fabricius and Galileo. There is no better example than Harvey who, as an Aristotelian sought out the method of the *Posterior Analytics* and the *Physics*—works which called for a dual role of experiment and reason.[2]

However, the significance of Padua should not blind us to the fact that most universities were conservative at the time and that their prime function was generally thought to be the preservation of the learning of the past rather than the quest for new knowledge through research.[3] This is surely in keeping with the spirit of the literary renaissance through which scholars were encouraged to seek out the lost treasures of antiquity rather than make new discoveries on their own. An excellent example of this may be taken from the field of medicine. Thomas Linacre (1460–1524), chief among the founders of the Royal College of Physicians in London (1518), had obtained his M.D. degree at Padua (*c.* 1492), and there he had diligently sought previously unknown Greek texts of the ancient medical authorities. These he published with his own translations.[4] Linacre, with many of his colleagues who were founders of the College, excelled in this typically Renaissance form of research. In England both the Elizabethan Statutes for Cambridge (1570) and the Laudian Code for Oxford (1636) maintained the authority of the ancients,[5] and this spirit also was carried over to professional societies. When Dr John Geynes had the temerity to suggest that Galen was not infallible (1559), he was forced to sign a recantation before being received back into the company of the Royal College of Physicians.[6] It was surely acceptable—even commendable—

8

for the humanist to criticize the vulgar annotations and emendations foisted on the works of the ancients by the Arabs and the Scholastic philosophers, but for many the original and pure texts seemed like an impregnable fortress of truth not to be added to or altered in any way.

This apparent—if not always actual—rejection of innovations could not help but result in a strong reaction in such an age of individualism. It is precisely this reaction which may be referred to as the first stage of the Scientific Revolution. Thus, at his examination for the m.a. in 1536, Petrus Ramus began his attack on Aristotle in defending the thesis 'Everything which Aristotle states is false.'[7] Similarly—in Italy—Telesio launched a strong attack on Medieval Aristotelianism in his Academy at Cosenza. Both Ramus and Telesio stressed the observation of Nature as a new foundation of knowledge rather than the constant repetition of the views of the ancients.[8] And as we move into the opening years of the new century we find an ever-increasing number of scholars calling for a new science based on observation and experiment. In his great experimental work on the magnet William Gilbert states that his natural philosophy 'is almost a new thing, unheard of before . . . . Therefore we do not at all quote the ancients and the Greeks as our supporters.'[9] And the Aristotelian, William Harvey, before discussing the circulation of the blood, states that 'I profess both to learn and to teach anatomy, not from books but from dissections; not from the positions of philosophers but from the fabric of nature.'[10]

And what of the formal training of the period? We do not have to look far to see distrust and rejection in the writings of many of these innovators. Descartes tells us that after completing the entire course of study at one of the most celebrated Schools of Europe, 'I found myself embarrassed with so many doubts and errors that it seemed to me that the effort to instruct myself had no effect other than the increasing of my own ignorance.'[11] Perhaps, he adds, the whole body of the sciences need not be reformed, 'but as regards all the opinions

9

which up to this time I had embraced, I thought I could not do better than endeavour once and for all to sweep them away, so that they might later on be replaced . . . .'[12]

If we turn to Francis Bacon, we find in the midst of his search for a new scientific method a critical reappraisal of the university system. After complaining of the low salaries allotted to faculty members, and after questioning the wisdom of concentrating studies on specific professions rather than the arts and sciences, he goes on:

Again, in the customs and institutions of schools, academies, colleges, and similar bodies destined for the abode of learned men and the cultivation of learning, everything is found adverse to the progress of science. For the lectures and exercises there are so ordered, that to think or speculate on anything out of the common way can hardly occur to any man. And if one or two have the boldness to use any liberty of judgment, they must undertake the task all by themselves; they can have no advantage from the company of others. And if they can endure this also, they will find their industry and largeness of mind no slight hindrance to their fortune. For the studies of men in these places are confined and as it were imprisoned in the writings of certain authors, from whom, if any man dissent he is straightway arraigned as a turbulent person and an innovator . . . . [The] . . . arts ands ciences should be like mines, where the noise of new works and further advances is heard on every side. But though the matter be so according to right reason, it is not so acted on in practice; and the points above mentioned in the administration and government of learning put a severe restraint upon the advancement of the sciences.[13]

In his utopian *New Atlantis* Bacon went on to describe 'Saloman's House' with its provisions for workshops, scientific instruments and laboratories—all of the requirements for the proper collection of scientific data according to Bacon's scheme for a new science. This was to become a major source of inspiration for the founders of the Royal Society of London. In short, when we look at Ramus, Telesio, Gilbert, Harvey, Descartes and Bacon—all of whom who have been referred to as founders of modern scientific method, we sense a disillusionment with the formal training of the period with its emphasis on the infallible truths of antiquity. And no matter what differences there might be between them in their own views on scientific method, they were agreed that in the future more attention must be paid to fresh observations and experiments.

And yet these views were not confined to that handful of great scholars such as these—men to whom we dedicate buildings and raise statues. If we shift our attention to the sixteenth-century iatrochemists, the Paracelsians and the alchemists, we find the same emphasis on new observations as a basis for a new science. We find with them *also* the same distrust of the traditional learning at the universities. The solution is to be a chemically orientated universe in which medicine and chemistry were all for practical purposes equated. Only on these new truths could a proper reform of education be founded. Let us look briefly at these few points.

First, the observational basis of the new science. Writing in 1538 Paracelsus stated 'I do not here write out of speculation, and theorie, but practically out of the light of Nature, and experience, lest I should burden you and make you weary with many words.'[14] The Paracelsian chemical physician is a man who is not afraid to work with his own hands. He is a pious man who praises God in his work and who lays aside all those vanities of his Galenist competitor and instead finds his delight in a knowledge of the fire while he learns the degrees of the science of alchemy.[15] The Paracelsian physician to the King of Denmark, Peter Severinus, wrote in 1571 that honest students of nature should sell their possesions and burn their books. With the proceeds they must buy sturdy clothes and set out to examine and observe everything with their own eyes. Above all, they should purchase coal, 'build furnaces, watch and operate with the fire without wearying. In this way and no other you will arrive at a knowledge of things and their properties.'[16] And R. Bostocke who effectively introduced Paracelsian thought to England in 1585 compared the traditional scholars who quoted Galen and Aristotle alone as the procedure for obtaining the doctorate with the new chemical physicians who try 'all things by fire whereby the vertue, nature, and propertie of each thing appeareth to the palpable and visible experience . . . .'[17] Bostocke goes on: 'He that listeth to leane to Bookes, let him learne of those Bookes which *Paracelsus* hath most Godly and learnedly expressed in his *Labyrinth*. In comparison of which al other Aucthorities in those matters are small or none.'[18] If we follow Bostocke's advice and turn to the *Labyrinthus Medicorum* of Paracelsus (1538) we find no written books discussed at all. Instead we are told to seek out God through nature. We must turn to the book of the heavens, the book of the elements, the book of man, the book of alchemy and the book of medicine. In effect, published books are nothing for science and knowledge must be based on experience and observation. 'Scientia enim est experientia.'[19]

Secondly, let me turn to the chemists' view of the state of formal education in their day. As Descartes and Bacon were appalled by the universities, so too were these chemical philosophers. The teaching of Paracelsus at the University of Basel in 1527 had been an unhappy experience remembered largely for his conflict with the students and faculty alike as well as his insistence on lecturing in Swiss-German and his burning of the *Canon* of Avicenna. The Paracelsian disillusionment with the schools is reflected throughout the sixteenth and the seventeenth centuries. Bostocke complained that although Paracelsian chemistry can be shown by experience to be true, its followers were in a desperate state since no one would give them a fair trial.

. . . in the scholes nothing may be received nor allowed that savoureth not of *Aristotle*, *Gallen*, *Avicen*, and other Ethnickes, whereby the yong beginners are either not acquainted with this doctrine, or els it is brought into hatred with them. And abrode likewise the *Galenists* be so armed and defended by the protection, priviledges and authorities of Princes, that nothing may bee received that agreeth not with their pleasures and doctrine . . . .[20]

Or we may compare Descartes' account of his education with that of van Helmont who was surely the most influential seventeenth-century iatrochemist:

At the age of seventeen I had finished my course in philosophy, and it was then that I noticed that nobody was admitted to the examination who was not masked in his gown and hood, as if the robes warrant scholarship. The professors made a laughing stock of the academic youth that was to be introduced to the arts and learning, and I could not help wondering at a sort of delirium in the behaviour of the professors, nay of everybody as much as the simplicity of the credulous youth. I retired into a deliberation in order to judge myself how much I was of a philosopher and had attained truth and science. I found myself inflated with letters and, as it were, naked as after partaking of the illicit apple—except for a proficiency in artificial wrangling. Then it dawned on me that I knew nothing and that what I knew was worthless. I did astronomy, logic and algebra for pleasure, as the other subjects nauseated me, and also the *Elements* of Euclid, which became particularly congenial to me since they contained the truth . . . . But I learned only vain eccentricities and a new revolution of the celestial bodies, and what seemed hardly worth the time and labours I had spent . . . . Having completed my course, I refused the title of Master of Arts since I knew nothing substantial, nothing true; unwilling to have myself made an arch-fool by the professors declaring me a Master of the Seven Arts, I who was not even a disciple yet. Seeking truth and science, though not their outward appearance, I withdrew from the university.[21]

There seemed no question of the need for a new experimental science and that the universities of the day were not providing this. But before discussing the reforms of the chemists, we must take up the question of what these men meant by chemistry and what they sought as their goal.[22] Above all, living in an intensely religious age they sought a true understanding of their Creator. Because of this, Aristotle and Galen could be—and were—severely attacked. Aristotle had been a heathen whose work had been condemned repeatedly in Church Councils, yet his works still formed the basis of philosophical thought at the universities. And for the Paracelsians, for whom medicine was a godly science second only to theology—and here witness *Ecclesiasticus* chapter 38 where the priestly office of the physician is described, and where medicine is referred to as a divine rather than a mundane science—for these men Galen was pictured primarily as a persecutor of Christians who had totally debased the study of man by his own slavish addiction to the Aristotelian corpus. In contrast, the work of Paracelsus in medicine was compared with that of Copernicus in Astronomy and with that of Luther, Melachthon, Zwingli and Calvin in theology. It is hardly surprising then that we should find these men continually referring to the two-book theory of knowledge. Man may obtain truth both through the Holy Scriptures or some mystical religious experience and through his diligent study of nature, God's book of Creation. Following the Hermetic tradition special emphasis was placed on the first chapter of *Genesis*. Here was a divine account of the formation of the world—an indisputable account. However, interpretations were not out of order, and for the Paracelsians the Creation was seen essentially as a divine *chemical* separation in which special emphasis was placed on the elements from which all other substances derive. Whether the particular author centred his discussion around the ancient Aristotelian elements or the newer Paracelsian principles, or even if he set up an independent element system of his own, he was able to think of this as a basis for all created substances as well as a cosmological system. Indeed, the significance placed on this Biblical account by the Paracelsians goes far to explain why there exists such a voluminous literature relating to element theory in the Renaissance, and why attacks on the Paracelsian principles should have been considered essential by opponents of the iatrochemists.[23]

Chemistry then had a divine significance. Since the Creation might be understood as a chemical process, it was thought that nature must continue to operate in chemical terms. It was the key to nature—all Created nature. Both earthly and heavenly phenomena were thought of as chemical processes and interpreted in this fashion. The scholar should go out to observe nature and collect specimens for study. These samples were then to be subjected to chemical analysis—that is, decomposition by heat. Aristotelians and Paracelsians alike believed that heat would reduce a substance to its elements. Chemistry then seemed to be a basic method of getting to fundamentals in nature.

However, there was more to the system than this. Most of these men were convinced of the truth of the macrocosm-microcosm relationship. The universe was conceived to be relatively small with all of its parts interconnected. In particular, man was seen as a small copy of the great world. The true physician will study all nature as he learns of man. Oswald Croll concentrated on this 'divine Analogy of this visible World and Man' in his influential *Basilica Chymica* of 1609. He affirmed that

Heaven and Earth are Mans parents, out of which Man last of all was created; he that knowes the parents, and can Anotomize them, hath attained the true knowledge of their child man, the most perfect creature in all his properties; because all things of the whole Universe meet in him as in the Centre, and the Anotomy of him in his Nature is the Anotomy of the whole world . . . .[24]

And while the ordinary philosopher might identify relationships between heaven and earth through outward signs, the chemical philosopher could learn far more through the use of his art and his analyses which might bring forth the hidden secrets of earthly bodies—and through analogy teach of the heavens. In the literature we find tables of correspondences between macrocosmic and microcosmic events. Robert Fludd gives a chemical description of the circulation of the blood which parallels the circular motion of the Sun in the great world,[25] and Joseph Quercetanus describes the running nose associated with a bad cold in terms of the formation of clouds and rains so that if we investigate the microcosmic phenomena we may hope to learn also of macrocosmic truths such as the source of winds, sleet and snow.[26] Glauber, one of the greatest of the practical chemists of the mid-seventeenth century, was quite willing to argue that since it had now been proved that the blood circulates in the body, we could hardly argue about the truth of such circulations in the macrocosm.[27] When reading this literature one must continually keep reminding oneself that such examples are always thought to be within the realm of chemistry.

In short, these men sought new observations and experiments no less than did authors whom we today honour as founders of the new science of the seventeenth century. Similarly they rejected the formal university training of the period in no uncertain terms. On these grounds they may be classed as 'moderns'. On the other hand, this science had incorporated in it an archaic world view which fundamentally separates it from man's approach to nature today. It is hardly surprising that Paracelsian thought should have had a natural appeal for physicians interested in a new science. It stressed medicine as the apex of natural philosophy, it bolstered its claims with the testimony of the Bible in an age when theology claimed the attention of every man, and it was openly experimental in its goal.

Evidence of the popularity of the chemical-Paracelsian approach to nature may be found in the ever-increasing number of books published on these topics in the seventeenth century. Even more convincing is the alarm we note in the works of men we might class as sounder scientists. We find Father Marin Mersenne disturbed about the new works of the chemists and seriously discussing the question whether alchemy should be considered an exact science.[28] We find Pierre Gassendi preparing a careful refutation of the works of the English physician-alchemist, Robert Fludd.[29] And we find Kepler engaged in a verbal duel with Fludd also—a debate which lasted for years.[30] As for the chemists themselves, we sense a spirit of success. Oswald Croll felt certain that Paracelsus' dream of overturning the ancient doctrines of the schools was imminent if not yet quite achieved. He pointed out that the courts of Europe did not lack competent Paracelsians, and he attributed their success to the truth of their chemical hypotheses, to the inherent progress of medical knowledge, and the elegant simplicity of the macrocosm-microcosm analogy.[31]

With the apparent success of their system, the problem of educational reform became much more important for the chemical philosophers. It is a topic referred to in many works —far more than we can discuss here. Let us choose only a few examples. The least promising of them all might appear to be the *Fama Fraternitatis* and the *Confessio* (1614, 1615). These are the texts which launched the Rosicrucian movement which to many may seem to represent the very essence of modern occultism. Reading these tracts in the framework of early seventeenth-century thought, however, a slightly different picture emerges. The *Fama Fraternitatis* in effect represents a call for a new learning.[32] Scholars today, the author complained, still pore over the atrophied learning taught in the schools—Aristotle, Galen and Porphyry—while instead they should be seeking a more perfect knowledge of the 'Son Jesus Christ and Nature'.[33] All truly learned men agree that the basis of natural philosophy is medicine and that this is a godly art.[34] The true secrets of medicine—and thus all science—had been learned by the founder of their order, Christian Rosenkreuz, and he had set them down in the secret books which are studied by the brethren of the order.[35] But great truths are to be found among those who are not members at all. In Germany today there are many learned magicians, physicians and philosophers—and such a one was the great Theophrastus Paracelsus whose works are concealed within the hidden vault of the Rosicrucians next to the special books of Christian Rosenkreuz.[36] This was indeed a neo-Paracelsian and an alchemical movement—and a missionary one at that. How much might be accomplished if the truly learned scholars of Europe united for the benefit of mankind? But, the author continues, where are we to find them? Surely they are not at the universities. They must declare themselves and join the Brotherhood in this reformation of learning. Accordingly one reads a plea in the *Fama Fraternitatis* addressed to all the learned scholars of Europe to examine their art 'and to declare their minde, either *Communicatio consilio*, or *singulatim* by Print'.[37] Both the *Fama* and the *Confessio* were to be published simultaneously in five languages so that no one could excuse himself and say that he had not seen the message—and while the Brothers refused at this time to give out their names or announce their meetings, they were willing to assure those who answered their call that their words would not go unnoticed.[38]

The *Fama* was actually published in four languages in nine editions between 1614 and 1617—and an English translation appeared in 1652.[39] Indeed, the response to this appeal must have been beyond the wildest dreams of the promoters. In the course of less than ten years several hundred books and tracts appeared debating the merits of this secret group, while major cities—such as Paris in 1623[40]—were visited by men who announced themselves as members of the Brotherhood and promised to show all of their secrets to those who wished to be initiated. In an account written in 1619 we read:

What a confusion among men followed the report of this thing, what a conflict among the learned, what an unrest and commotion of imposters and swindlers, it is needless to say. There is just this one thing which we would like to add, that there were some who in this blind terror wished to have their old, out-of-date, and falsified affairs entirely retained and defended with force. Some hastened to surrender the strength of their opinions; and after they had made accusation against the severest yoke of their servitude, hastened to reach out after freedom.[41]

In the rash of utopian schemes of this era we find the *Christianopolis* of Johann Valentin Andreae (1619), the man who may have been the author of the *Fama*. While this work is little known today, it bears striking similarities to the *New Atlantis* of Bacon which was written shortly after, and its importance for seventeenth-century science seems even greater than this.[42] Andreae's close connection with Comenius who was in turn the friend of Hartilb, Dury and Petty connects him with the background which culminated in the foundation of the Royal Society.[43] In the *Christianopolis* Andreae deplores the decay of learning and religion and suggests that a proper community be formed—open to all of good intent and character.[44] We need not discuss this model in detail. We need only note that here again we see a reaction against tradition in the spirit of the contemporary chemical philosophers. The citizens of *Christianopolis*, for instance, have a large library, but they use it little and concentrate only on the most thorough books. The meaning is clear:

The highest authority among them is that of sacred literature, that is, of the Divine Book; and this is the prize which they recognize as conceded by divine gift to men and of inexhaustible mysteries; almost everything else they consider of comparatively little value . . . .[45]

The scholars here prefer to get their knowledge directly from the Book of Nature. A study of nature brings about a greater understanding of our Creator for 'a close examination of the earth will bring about a proper appreciation of the heavens, and when the value of the heavens has been found, there will be a contempt of the earth'.[46]

Accordingly we are more than justified in looking at the centre for such studies, the laboratory. As we might expect, it is a chemistry laboratory fitted out with the most complete equipment and it is here that 'the properties of metals, minerals, and vegetables, and even the life of animals are examined, purified, increased, and united, for the use of the human race and in the interests of health'.[47] More important, however, is the fact that it is here that 'the sky and the earth are married together', and it is here that the 'divine mysteries impressed upon the land are discovered'.[48] These are clearly references to

19

the macrocosm-microcosm analogy and the doctrine of signatures.

The importance of chemistry for Andreae becomes even more evident when one compares it with his treatment of other sciences. In the hall of physics we do not find basic studies of motion. Instead, one views natural history scenes painted on the walls—views of the sky, the planets, animals and plants. The visitor may examine rare gems and minerals, poisons and antidotes and all sorts of things beneficial and injurious to the several organs of man's body.[49] Here the study of mathematics rises above vulgar arithmetic and geometry to the mystical numerical harmonies of the heavens —a subject known to the Pythagoreans of old.[50] The relation of heaven and earth is everywhere stressed, and for this reason astrology is raised to its proper place. Andreae states that

he who does not know the value of astrology in human affairs, or who foolishly denies it, I would wish that he would have to dig in the earth, cultivate and work the fields, for as long a time as possible, in unfavourable weather.[51]

The implication is clear. A new learning is required, and if it cannot be accommodated to the current university system, a separate academy or college must be founded. Andreae's proposals could have been seconded by any of the chemical philosophers.

We have yet to examine any schemes for educational reform proposed by the chemists themselves. Let me confine my remarks to two contemporaries, Robert Fludd and Jean Baptiste van Helmont. These men represent two extremes within the same camp. We must take up Fludd first since his views are not too far removed from the Hermetic approach of the *Christianopolis* and the *Fama Fraternitatis*. Indeed, Fludd is one of those who answered the plea to the learned community of Europe printed in this Rosicrucian text. Fludd was a knight, a man of substance who had been educated at Oxford and then had spent years touring the educational centres on the continent.[52] After his return home he was admitted as a fellow of the Royal College of Physicians (1609) where he became the friend of William Harvey and Mark Ridley. Yet, this man was far from being a 'modern'. Fludd wrote with sincerity about the mystical alchemical Creation, and he considered kabbalistic analysis to be a proper method in the scholar's search for divine secrets. His publications were controversial in tone and widely discussed.

When the *Fama Fraternitatis* appeared Fludd had as yet put nothing into print. However, as a Hermeticist of long standing he was attracted by its message, and when another noted chemist, Andreas Libavius, attacked the Rosicrucians, Fludd rapidly penned a reply in which he made a plea for a new learning.[53] Nowhere did he feel that there was more useless knowledge being spewed forth than at the universities. These were the strongholds of the classical scholars who felt that truth might only be found in the writings of the ancients. One could only shudder at their reliance on Aristotle in philosophy and Galen in medicine, for in the ancient world none set forth doctrines more opposed to Christianity than they. Aristotle and Galen had been heathens, and their followers were no better. The universities must be reformed so that the divine light of Christian teachings could flourish. In a chapter on natural philosophy, medicine and alchemy, Fludd tells us that although innumerable authors have written on natural philosophy, they have presented to us only a shade of truth.[54] They fill whole volumes with definitions, descriptions and divisions, and they drone on about the four causes. They lecture on motion, the continuum, the contiguum—of termini, loci, of space, vacua, time and number. This for them is the basis of physics. They go forth then to generation, corruption and descriptive accounts of the heavens and the universe.[55] 'But, good God, how superficial and equivocal this all is.'[56] It is impossible for anyone to attain the highest knowledge of natural philosophy without a thorough training in the occult sciences.[57]

21

One must turn first to medicine, the most perfect science of all. There should not be understood by this word a simple description of diseases and the workings of the human body. Rather, it is the very basis on which natural philosophy must rest.[58] Our knowledge of the microcosm will teach us of the great world, and this, in turn, will lead us to our Creator. Similarly, the more we learn of the universe, the more we will be rewarded with a perfect knowledge of ourselves. Obviously the medical schools must change their emphasis.

The art of alchemy from which we might expect so much in reality aids us little. Alchemists like Andreas Libavius do not call for a reform, but rather describe calcination, separation, conjunction, putrefaction, and all the other operations by which men are deceived and persuaded to part with their money.[59] True chemical authors are of a different sort and write of occult secrets.[60] The science is a system of nature—not a refuge for gold-makers.

Real wisdom may be found in the writings of the natural magicians, men who are in truth mathematicians.[61] But if Fludd was distressed about the state of natural philosophy, medicine and alchemy in his day, he felt even more strongly about mathematics.[62] The texts of arithmetic were filled with definitions, principles and discussions of theoretical operations. We learn of addition, subtraction, multiplication, division, golden numbers, fractions, square roots and the extraction of cubes.[63] But of the old wisdom and doctrines of the Pythagoreans the Arithmeticians pay little heed. These men of old were silent about the arcane arithmetic, and like mystical alchemists they hid their profound mysteries from the vulgar crowd.[64]

And, Fludd continues, if we go beyond arithmetic to other mathematical studies, we find the same superficiality. Those who are really profound in these arts are considered impious and magicians.[65] Indeed, we can learn more from the silent wisdom of the Pythagoreans than we can from the useless books of the philosophers. The disciples of Pythagoras reached a certainty of belief in God through their profound study of numbers and their ratios. In the same way, we may learn of the unity of God in trinity,—and more, of the very construction of the world.[66]

We need not continue much farther. For Fludd, music really deals with the joining of the elements, the proportions of light and weight in the stars, and their influence on the terrestrial world—in short, it relates primarily to the celestial harmony.[67] Geometry really concerns other hidden secrets. Surely the greatest geometer of all time was Archimedes, and Fludd turns no less resolutely than the mechanical philosophers to Archimedes as a guide to a new science. Why?—because of his wondrous machines.[68] In effect, Archimedes is to be considered as the archetype of the perfect natural magician.

Fludd represents an extreme case of the mystical alchemist who sought hidden secrets in the writings of the sages of old. When pressed to do so he could debate over the problems of experimental method as a modern, but experiment for him always to be subservient to the revealed truths of religion, and Holy Writ would always carry greater weight than laboratory results. It is significant that his whole scheme was chemically orientated, and that for him only reforms of the most drastic sort could save the university system.

No greater contrast to Fludd could be found among the chemical philosophers than Jean Baptiste van Helmont, a man who was appalled by the mystical approach of Fludd and who deplored the use of the macrocosm-microcosm analogy in nature. He writes that

The name of Microcosm or little World is Poetical, heathenish, and metaphorical, but not natural, or true. It is likewise a Phantastical, hypochondriacal and mad thing, to have brought all the properties, and species of the universe into man, and the art of healing.[69]

We have already noted his disillusion with his training at school. Yet, in his reaction to this he had immersed himself in the Hermetic and Paracelsian texts as the only major alternative. Consequently, even though he later rejected elements of Paracelsism, here too we find many of the same currents of thought. No less than Paracelsus or Fludd, van Helmont insists that man may safely rely on *Genesis* as a basis for physical truth. When he rejects the Aristotelian four elements it is on this basis—they are not mentioned in the Creation account.[70] He could write almost in rapture:

I praise by bountiful God, who hath called me into the Art of the fire, out of the dregs of other professsion. For truly Chymistry, hath its principles not gotten by discourses, but those which are known by nature and evident by the fire: and it prepares the understanding to pierce the secrets of nature, and causeth a further searching out in nature, than all other Sciences being put together: and it pierceth even unto the utmost depths of real truth . . . .[71]

Van Helmont's disillusion with the teachings of the schools and their emphasis on Galen and Aristotle may be traced at least in part to his views on mathematics. As a student he had laboured over logic and the principles of Euclid which seemed to contain truth, but when he sought theorems and axioms in medicine, he found this quest quite hopeless.[72] Medicine is a divine rather than an ordinary science because Scripture teaches us that the Lord created physicians.[73] Rational, mathematically inspired investigations may aid us in the study of physics, but not the chief goal of natural philosophy, medicine, for to 'understand and favour these things from the spring or first cause is granted to none without the special favour of Christ the Lord'.[74] Neither the cosmic mathematics of the mystical alchemists nor the logical-mathematical method of the Galenist may help us here.

And if we turn to the mathematical physics of the Aristotelians we will do no better than before. The teachings of Aristotle dominate our schools and turn the minds of our scholars into erroneous paths. Yet, if we examine his writings we shall unmask him. It is true that Aristotle had a most persuasive manner in arranging rules and maxims. In a later—and more lazy—age students willingly accepted these as a guide to truth while they prostrated themselves to him almost in worship.[75] But the evidence shows, van Helmont adds, that Aristotle surely had little ability as a scientific observer. Therefore he relied heavily on mathematics in preparing his vast system simply because he was 'far more skilful in this, than in Nature'. And since this man had subdued nature under the rules of mathematics, our scholars are now cursed with an improper approach to science.[76]

Thoroughly sickened by the logical-mathematical methods of the schools, van Helmont could only hope—as did so many of his generation—for a new learning based upon a total reform. In his early work on the magnetic cure of wounds we see van Helmont's acceptance of natural magic as 'the most profound inbred knowledge of things',[77] and if he was to reject the most occult aspects of Paracelsian thought, we still find his emphasis on experiment and chemistry in his scheme for educational reform. No longer should young men waste their youth on Aristotelian thought—a term which for van Helmont could almost be equated with mathematical thought. In a seven-year programme they may devote three years to arithmetic, mathematical science, the *Elements* of Euclid, and geography (which is to include the study of seas, rivers, springs, mountains, and minerals). In addition—in this three-year period—they are to study the properties and customs of nations, waters, plants, living creatures, and the ring and astrolabe. Only then, van Helmont continues

. . . let them come to the Study of Nature, let them learn to know and seperate the first Beginnings of Bodies. I say, by working, to have known their fixedness, volatility or swiftness, with their seperation, life, death, interchangeable course, defects, alteration, weakness, corruption, transplanting, solution, coagulation or co-thickening, resolving. Let the History of extractions, dividings, conjoynings, ripenesses, promotions, hinderances, consequences, lastly, of losse and profit, be added. Let them also be taught, the Beginnings of Seeds, Ferments, Spirits, and Tinctures, with every flowing, digesting, changing, motion, and disturbance of things to be altered.

25

And all those things, not indeed by a naked description of discourse, but by handicraft demonstration of the fire. For truly nature measureth her works by distilling, moystening, drying, calcining, resolving, plainly by the same meanes, whereby glasses do accomplish those same operations. And so the Artificer, by changing the operations of nature, obtains the properties and knowledge of the same. For however natural a wit, and sharpness of judgement the Philosopher may have, yet he is never admitted to the Root, or radical knowledge of natural things, without the fire. And so every one is deluded with a thousand thoughts or doubts, the which he unfoldeth not to himself, but by the help of the fire. Therefore I confess, nothing doth more fully bring a man that is greedy of knowing, to the knowledge of all things knowable, than the fire. Therefore a young man at length, returning out of these Schooles, truly it is a wonder to see, how much he shall ascend above the Phylosophers of the University, and the vain reasoning of the Schooles.[78]

This is the Helmontian-iatrochemical programme for a reform in education. Students are to spend the bulk of their time learning the true facts of nature through observation and chemical operations with the fire.

The search for a new learning based on chemistry and the occult sciences reached its peak in the middle decades of the seventeenth century. England in the Civil War period was particularly susceptible to such speculations. Indicative of the increased interest is the case of an astrological work written in 1608 which was withheld from the printer until 1650. In the preface we are told that men need no longer be afraid to set forth such views. They are now received favourably by a large segment of the scholarly public and they are well defended by the members of the learned London Society of Astrologers.[79] Paracelsian and Helmontian chemistry shared in the growing interest in all schemes for a new science. In the decade of the 1650s more Paracelsian and mystical chemical works were translated than in the entire century before 1650. At the same time, and with good reason, there was also a greater interest in the Rosicrucian movement. John Heydon gave his books such titles as *A New Method of Rosie Crucian Physicke* (1658) and *The Rosie Crucian Infallible Axiomata* (1660), while Eugenius Philalethes (Thomas Vaughn) prepared a lengthy introduction to a translation of *The Fame and Confession of the Fraternity of R:C:* (1652). The connection was quite clear for George Hakewill in his survey of scholarship. He praised the '*Chimiques*, *Hermetiques*, or Paracelsians (& a branch of them as I conceive is the order Roseae Crucis)'.[80]

The frustration of the chemists is shown by their attempt to break the control over physicians exerted by the Royal College. This they hoped to do by forming rival societies of chemical physicians. And if the schools were not open to their teachings, let a public test be made to prove the truth of their assertions. Listen to van Helmont:

Oh ye Schooles .... Let us take out of the hospitals, out of the Camps, or from elsewhere, 200, or 500 poor People, that have Fevers, Pleurisies, &c. Let us divide them into halfes, let us cast lots, that one halfe of them may fall to my share, and the other to yours; ... we shall see how many Funerals both of us shall have: *But* let the reward of the contention or wager, be 300 Florens, deposited on both sides: Here your business is decided.[81]

This challenge was repeated continuously by the chemical physicians of the Commonwealth and the early Restoration.[82]

The situation at the universities remained the most galling thorn of all. Surely the teaching of scientific subjects was not a major strength of Cambridge or Oxford at this time, and for the chemists the problem seemed intolerable. In his *Art of Distillation* (1650) John French spoke of alchemy 'which is more noble than all the other six Arts and Sciences ...':

This is that true natural philosophy which most accurately anatomizeth Nature and natural things, and ocularly demonstrates the principles and operations of them: that empty natural philosophy which is read in the universities, is scarce the meanest hand-maid to this Queen of Arts. It is a pity there is such great encouragement for many empty, and unprofitable Arts, and none for this, and such like ingenuities, which if promoted would render an University far more flourishing, than the former. I once read or heard, of a famous University beyond Sea, that was faln into decay, through what cause I know not: but there was a general counsel held by the learned, how to restore it to its primitive glory: The *Medium* at last agreed upon, was the promoting of Alchymie, and encouraging the Artists themselves: But I never expect to see such rational actings in this nation till shadows vanish, substances flourish, and truth prevail . . . .[83]

French may have practically given up hope for such rational behaviour in England, but not others. In an inflammatory work we find Noah Briggs—called a 'Psittacum Helmontii' by an opponent—demanding chemistry as the proper model science for students at the universities (1651),[84] while John Hall, addressing *An Humble Motion to the Parliament of England Concerning the Advancement of Learning: and Reformation of the Universities* (1649), asked: 'Where have we any thing to do with Chimistry, which hath snatcht the keyes of Nature from the other Sects of Philosophy, by her multiplied real experiences?'[85]

These pleas for a reform of higher education came into direct conflict with the proponents of the mechanical philosophy. This may be seen in the famous debate between John Webster and Seth Ward in 1654. John Webster (1610–82) represents the reforming chemists of the era, men who were seeking an educational reform based on Christian principles. To them the works of Fludd and the Paracelsians seemed to be the answer to the stale learning taught at the universities. Webster is a typical example of this school.[86] Attracted early both to the study of nature and religion, he studied chemistry under the Hungarian alchemist, John Hunyades, (*c.* 1632)[87] and was ordained as a minister shortly after that date. With his Puritan sympathies he served as a surgeon and chaplain with the Parliamentary army during the Civil War. By 1648 his reaction against the established Church had forced him to become a nonconformist. Although most of his writings are on religious topics, his important scientific work, the *Metallographia*, went through two editions (1661, 1671).

John Webster's strong views on the reform of the universities had prompted him to write his *Academiarum Examen* in 1654. This work, 'offered to the judgements of all those that love the proficiencie of Arts and Sciences and the advancement of Learning', is in reality a call for reform in terms similar to those advocated by Robert Fludd in his *Tractatus Apologeticus* of 1617. Reacting against the sterile, and to his mind atheistic writings of Aristotle taught at the universities, Webster spoke instead of the 'mysterious and divinely-inspired *Teutonick* [Jacob Boehme], and . . . the highly illuminated fraternity of the Rosie Crosse'.[88] As the Paracelsians teach, true Christian knowledge of Nature will be taught best by ocular demonstrations learned by putting 'hands to the coals and furnace'.[89] In this way we shall learn the importance of the three principles, while we must also seek out the secrets of natural magic and 'Cabalistick Science'.[90] In general if we are properly to reform our knowledge on Christian principles we must surely seek to build up tables of axioms as Bacon has suggested, but we must—as true Christians—seek a knowledge of nature

that is grounded upon sensible, rational, experimentall, and Scripture principles: and such a compleat piece in the most particulars of all human learning (though many vainly and falsely imagine there is no such perfect work to be found) is the elaborate writings of that profoundly learned man Dr. *Fludd*, than which for all the particulars before mentioned (notwithstanding the ignorance and envy of all opposers) the world never had a more rare, experimental and perfect piece.[91]

In addition, the new philosophers are told to avoid Aristotle and to turn to the works of Ficino, Plato, Gilbert and Hermes Trismegistus—as interpreted by the Paracelsians. As might be expected, experimental chemistry is to be the new key to nature, and the medicine of Paracelsus and van Helmont is to replace that of Galen.[92]

John Webster's book was scathingly attacked by Seth Ward (1617–89), who, in contrast with Webster, is today revered as one of the founders of modern science. Educated at Cambridge, Ward had early showed an aptitude in mathematics and astronomy. Although he had been chosen mathematical lecturer at Cambridge in 1643, Ward was deprived of his fellowship in 1644 because of his staunch adherence to the established Church. After a period of wandering, he moved to Oxford (1647) where he was appointed Savilian professor of astronomy. This was a period when Oxford was rapidly becoming a centre for the new science. In addition to Ward, Robert Boyle, Thomas Willis, Jonathon Goddard, John Wallis and John Wilkins were at Oxford in the early 1650s. As a group, these men formed the nucleus of the 'Philosophical Society of Oxford' and this, in turn, was a forerunner of the Royal Society of London. It is little wonder with these brilliant scholars about him that Seth Ward took exception to John Webster's attack on the universities.

Perhaps forgetting that at Oxford the lectures in astronomy had fallen into neglect before his appointment in 1649, Ward in his *Vindiciae Academiarum* emphasised the high level of scientific work at the universities—and how improper and inconsistent Webster's suggestions really were. The prelude to the attack may be seen in John Wilkins' introduction. Speaking of Webster, he states:

the man doth give me the freest prospect of his depth and braine, in that canting Discourse about the language of nature, wherein he doth assent unto the highly illuminated fraternity of the *Rosy-crucians*. In his large enconiums upon *Jacob Behem*, in that reverence which he professes to iudiciall Astrologie, which may sufficently convince what a kind of credulous Reformer he is like to prove.[93]

Ward—in a point-for-point rebuttal of Webster's work—devotes much space to the state of mathematics at the universities which he admits could be bettered, but hardly in the fashion proposed by his adversary. He questions the wisdom of completely discarding the Galenic medicine, and he argues that Webster is unfair in his accusation that Chemistry is unknown at the universities—surely the group at Oxford included men who were active in this field.[94] And if Webster pointed to the writings of Francis Bacon as basic for a reform of nature, Seth Ward could hardly agree more—but, how does Webster follow up this suggestion? He says, indeed, that

The second Remedy is, *That some Physicall Learning may be brought into the Schooles, that is grounded upon sensible, Rationall, Experimentall and Scripture Principles, an such an Author is Dr Fludd; then which for all the particulars, the World never had a more perfect piece.*[95]

Surely this is too much! 'How little trust there is in villainous man!' Although a moment before he had recommended Francis Bacon 'for the way of strict and accurate induction', now he

is fallen into the mysticall way of the *Cabala*, and numbers formall: there are not two waies in the whole World more opposite, than those of L. *Verulam* and D. *Fludd*, the one founded upon experiment, the other upon mysticall Ideal reasons; even now he was for him, now he is for this, and all this in the twinkling of an eye, O the celerity of the change and motion of the Wind.[96]

And if he turns in Philosophy to Plato, Democritus, Epicurus, Philolaus and Gilbert, why should there be any need in this, for 'if *De Fluctibus* be so perfect, what need we go any farther?'[97]

The Webster-Ward debate is often misunderstood. John Webster has been accused of rashly proposing an odd mixture of science and superstition—chemistry and magic—as a basis for university reform.[98] To this Ward is praised for reacting in righteous anger. But as we have seen, Webster's view of chemistry is not what we mean by modern scientific chemistry. And his magic is far from the black art the word may imply to us. Here he would have agreed with Bacon that the term Natural Magic had long been misapplied, and that it really signified 'Natural Wisdom, or Natural Prudence . . . purged from vanity and superstition'.[99]

31

And if we find an incorrect interpretation of the work of Webster and Ward in microcosm, perhaps there exists the same error in macrocosm in our understanding of the Scientific Revolution. Today it would seem that it was primarily the development of physics which resulted in our new age of science. This future development, however, was far from evident to all natural philosophers who lived in the sixteenth and early seventeenth centuries, and for this reason it need not delay us unduly here. In reality Webster had been able to speak in terms which were not so out of joint with the times as we might think. It was an age when the widely respected van Helmont could ridicule the increasing interest in mathematical abstraction no less than he could the mystical cosmic mathematics of Fludd. It was an age in which we see Kepler immersed in geometrical mysticism, and one in which we find Galileo fascinated by the relationships of numbers. We see even the great Newton willing to devote an incredible amount of his time to the study of alchemical authors.

In short, the Scientific Revolution was not simply the forward march of a new experimental method coupled with the powerful tool of mathematical abstraction. For some the two were incompatible and the growing predominance of mathematical abstraction could be interpreted as a step backward—a step away from a truly experimental study of nature. For these men the purely experimental studies of chemistry seemed the best answer to the training of the logic-ridden universities. We have seen this in the glorification of Paracelsus in the Rosicrucian texts and in the chemically orientated laboratory science of the utopian *Christianopolis*. And with the two chemists we have touched on, Fludd and van Helmont, we note a similar hope for educational change— a hope dominated by their belief that chemistry was the true key to nature.

I would close by saying that I do believe that there was a chemical dream in the Renaissance—it was a search for our Creator through his created work by chemical investigations and analogies. Its appeal was widespread and its followers were sincere men. If today we find their efforts judged harshly, it is because we now live in a far different age. Their goals were not too dissimilar from those of the mechanical philosophers. To replace an over-emphasis on Greek authority they called for a new view of nature based on new observations and new experiments—and they recognised the need for educational reform so that students could benefit from this new learning. But if we are to understand these chemical philosophers we must be willing to look at the world not from our twentieth-century vantage-point, but from the viewpoint of their—now archaic—cosmology. When we have done this perhaps we will have moved one step closer to a true understanding of the birth of modern science.

NOTES

1. John Herman Randall, Jr., 'The Development of Scientific Method in the School of Padua', *Journal of the History of Ideas*, i (1940), 177–206.

2. The case of Harvey is described in detail by Walter Pagel, *William Harvey's Biological Ideas. Selected Aspects and Historical Background* (Basel/New York: S. Karger, 1967), pp. 28–47. On method the reader is also referred to the pertinent sections in W. P. D. Wightman, *Science and the Renaissance* (2 vols., Aberdeen U.P., 1962). See also Wightman's views as expounded in his paper 'Method and Myth in Seventeenth Century Biology' read at the Summer Meeting of the British Society for the History of Science, July 7, 1967. No attempt will be made in the present paper to assess the role played by Aristotelian thought on science in the sixteenth and seventeenth centuries.

3. See for instance Phyllis Allen, 'Scientific Studies in the English Universities of the 17th Century', *Journal of the History of Ideas*, x, 219–53; William T. Costello, s.j., *The Scholastic Curriculum at Early Seventeenth-Century Cambridge* (Cambridge, Mass.: Harvard U.P., 1958), p. 9; Mark H. Curtis, *Oxford and Cambridge in Transition* 1558–1642 (Oxford: Clarendon Press, 1959), p. 227.

4. William Munk, M.D., F.S.A., *The Roll of the Royal College of Physicians of London* . . . (2nd ed., 3 vols., London: Published by the College, 1878), i, pp. 12–21.

5. Allen, *op. cit.*, 219.

6. W. S. C. Copeman, *Doctors and Disease in Tudor Times* (London, 1960), p. 36.

7. R. Hooykaas, *Humanisme, Science et Réforme* (Leyden: E. J. Brill, 1958), p. 7.

8. On Telesio see Francesco Fiorentino, *B. Telesio, ossia studi storici su l'idea della natura nel risorgimento Italiano* (2 vols., Firenze, 1872–4). Telesio's main theoretical work on nature was the *De natura iuxta propria principia liber primus et secundus* (Rome: Antonium Bladum, 1565).

9. William Gilbert, *On the Loadstone and Magnetic Bodies, and on The Great Magnet the Earth*, trans. P. Fleury Mottelay (London: Bernard Quaritch, 1893), p. 1.

10. William Harvey, *The Works of William Harvey, M.D.*, trans., and with a life by Robert Willis, M.D. (London: Sydenham Society, 1847), p. 7.

11. R. Descartes, *Discours de la Méthode*, Introduction and notes by Étienne Gilson (Paris: J. Vrin, 1964), p. 49.

12. *Ibid.*, p. 62.

13. *The Philosophical Works of Francis Bacon . . . Reprinted from the Texts and Translations*, notes and prefaces of Ellis and Spedding, ed. with an Introduction by John M. Robertson (London: Routledge and Sons; New York: Dutton, 1905), p. 286 from the *Novum Organum*. See also *On the Advancement of Learning*, p. 76.

14. Paracelsus, *Of the Nature of Things* in Michael Sendivogius, *A New Light of Alchymy* (London: A. Clark for Tho. Williams, 1674), sig. L8 recto.

15. *Ibid.*, pp. 252–3.

16. Petrus Severinus, *Idea Medicinae Philosophicae* (3rd ed., Hagae Comitis, p. 39.

17. R. Bostocke, Esq., *The difference betwene the auncient Physicke . . . and the latter Phisicke* (London, Robert Walley, 1585), sig. Dv verso.

18. *Ibid.*, sig. Dvi recto.

19. Paracelsus, *Labyrinthus Medicorum* in *Opera Omnia* (3 vols., Geneva: Joan. Antonii & Samuelis De Tournes, 1658), i, pp. 264–88 (275).

20. Bostocke, *op. cit.*, sig. Fiii recto.

21. Walter Pagel's translation found in his 'The Reaction to Aristotle in Seventeenth-Century Biological Thought' in *Science, Medicine and History. Essays on the Evolution of Scientific Thought and Medical Practice written in honour of Charles Singer* (2 vols., London/New York/Toronto: Oxford U.P., 1953), i, pp. 489–509 (491–2). A seventeenth-century translation will be found in John Baptista van Helmont, *Oriatrike or Physick Refined*, trans. J. Chandler (London: Lodowick Loyd, 1662), pp. 11–12.

22. The following account is based primarily on the present author's 'Renaissance Chemistry and the Work of Robert Fludd' in *Alchemy and Chemistry in the Seventeenth Century* (Los Angeles: William Andrews Clark Memorial Library, 1966), pp. 3–29. This essay was printed in slightly revised form in *Ambix*, xiv (1967), 42–59. See also Allen G. Debus, *The English Paracelsians* (London: Oldbourne Press, 1965; New York: Franklin Watts, 1966).

23. See Allen G. Debus, 'Fire Analysis and the Elements in the Sixteenth and Seventeenth Centuries', *Annals of Science*, xxiii (1967), 127–47.

24. Osw. Crollius, *Discovering the Great and Deep Mysteries of Nature* in *Philosophy Reformed and Improved*, trans. H. Pinnell (London, 1657), p. 24.

25. Allen G. Debus, 'Robert Fludd and the Circulation of the Blood', *Journal of the History of Medicine*, xvi (1961), 374–93.

35

26. Joseph Du Chesne, *Traicté de la Matière, Preparation et excellente vertu de la Médecine balsamique des Anciens Philosophes* (Paris, 1626—first Latin edition, 1603), p. 183.

27. J. R. Glauber, *Works*, trans. Christopher Packe (London, 1689), pp. 248f.

28. P. Marin Mersenne, *La Vérité des Sciences Contre Les Septiques ou Pyrrhoniens* (Paris, 1625). This is discussed by R. Lenoble, *Mersenne ou la naissance du mécanisme* (Paris, 1943).

29. Pierre Gassendi, *Epistolica exercitatio in qua principia philosophiae Roberti Fluddi, medici, reteguntur, et ad recentes illius libras adversus R. P. F. Marinum Mersennum . . . respondetur* (Paris, 1630).

30. The Kepler-Fludd debate has been described in part by Wolfgang Pauli in 'The Influence of Archetypal Ideas on the Scientific Theories Theories of Kepler', in C. G. Jung and W. Pauli, *The Interpretation of Nature and the Psyche*, trans. P. Silz (New York, 1955).

31. Crollius, *op. cit.*, pp. 142–7.

32. Frances A. Yates touches on the Rosicrucian movement and early modern science in her 'The Hermetic Tradition in the Renaissance' in *The Johns Hopkins Seminars in the Humanities in the Renaissance*, ed. ed. Stapleton (Baltimore: The Johns Hopkins U.P., in press).

33. *The Fame and Confession of the Fraternity of R:C: Commonly, of the Rosie Cross. With a Praeface annexed thereto, and a short Declaration of their Physicall Work*. By Eugenius Philalethes (Thomas Vaughn) (London: J. M. for Giles Calvert, 1652), pp. 1–2.

34. *Ibid.*, p. 36.

35. *Ibid.*, p. 5.

36. *Ibid.*, p. 10.

37. *Ibid.*, p. 31.

38. *Ibid.*

39. A listing—and discussion—of the early Rosicrucian texts will be found in F. Leigh Gardner, *A Catalogue Raisonné of Works on the Occult Sciences, vol. i Rosicrucian Books*, intro. Dr William Wynn Westcott (2nd ed., Privately Printed, 1923). Here on pages 4–6, 21 (items 23–29, 144) will be found a discussion and description of the first editions of the *Fama Fraternitatis* and the *Confessio* (1614–1615). Other early editions are listed in Johann Valentin Andreae, *Christianopolis. An Ideal State of the Seventeenth Century*, trans. with an Historical introduction by Felix Emil Held (New York: Oxford U.P., 1916), p. 11.

40. See the article on Michael Maier in the *Biographie Universelle* (Paris: Madame C. Desplaces, n.d.), xxvi, p. 114.

41. *Andreae, op. cit.*, pp. 137–8.

42. *Ibid.*, pp. 53ff.
43. The influence of the *Christianopolis* has been touched on briefly by W. H. G. Armytage, 'The Early Utopists and Science in England', *Annals of Science*, xii, (1956), 247–54. Held has discussed the relation of the *Christianopolis* to the founding of the Royal Society in his edition of the text (cited above, note 39), pp. 100–25. Considerable attention is paid to this work by Margery Purver, *The Royal Society: Concept and Creation* (London: Routledge and Kegan Paul, 1967). See also G. H. Turnbull, *Hartlib, Drury and Comenius. Gleanings from Hartlib's Papers* (Liverpool: University of Liverpool Press, 1947), and H. R. Trevor-Roper, 'Three Foreigners and the Philosophy of the English Revolution', *Encounter*, xiv (1960), 3–20.
44. Andreae, *op. cit.*, p. 29.
45. *Ibid.*, p. 191.
46. *Ibid.*, p. 187.
47. *Ibid.*, pp. 196–7.
48. *Ibid.*
49. *Ibid.*, p. 200.
50. *Ibid.*, pp. 221-3.
51. *Ibid.*, p. 228.
52. In addition to the works cited in note 22, see J. B. Craven, *Doctor Robert Fludd* (Kirkwall, 1902) and C. H. Josten, 'Truth's Golden Harrow: An Unpublished Alchemical Treatise of Robert Fludd in the Bodleian Library', *Ambix*, iii (1949), 91–150.
53. Robert Fludd, *Apologia Compendiaris Fraternitatem de Rosea Cruce Suspicionis et Infamiae Maculis Aspersam, Veritatis quasi Fluctibus abluens et abstergens* (Leiden, 1616). This short pamphlet was greatly enlarged the following year as the *Tractatus Apologeticus Integritatem Societatis De Rosea Cruce defendens* (Lugduni Batavorum, 1617). This is the edition which will be cited here. A German translation of this text appeared as *Schutzschrift für die Aechtheit der Rosenkreutzergesellschaft . . . übersetzt von Ada Mah Booz* (Leipzig, 1782).
54. *Tractatus Apologeticus*, p. 91.
55. *Ibid.*, pp. 91-3.
56. *Ibid.*, p. 93.
57. *Ibid.*
58. *Ibid.*, p. 89.
59. *Ibid.*, pp. 100–1.
60. *Ibid.*, p. 101.
61. *Ibid.*, p. 23.
62. *Ibid.*, p. 103.
63. *Ibid.*

64. *Ibid.*, p. 105.

65. *Ibid.*

66. *Ibid.*, p. 107. Fludd would have rejoiced in the publication in the same year of a book on mystical geometry by Fortunatus, *Decas Elementorum Mysticae Geometrae Quibus Praecipua Divinitatus Arcana Explicantur* (Padua: Peter Paul Tozzi, 1617). I am indebted to Walter Pagel for this reference.

67. *Tractatus Apologeticus*, p. 109.

68. *Ibid.*, p. 114.

69. van Helmont, *Oriatrike*, p. 323.

70. *Ibid.*, p. 48.

71. *Ibid.*, p. 462.

72. *Ibid.*, p. 13.

73. *Ibid.*, p. 4.

74. *Ibid.*, p. 4.

75. *Ibid.*, p. 33.

76. *Ibid.*

77. van Helmont, *Oriatrike*, p. 784.

78. *Ibid.*, p. 45.

79. Sir Christopher Heydon, *An Astrological Discourse, Manifestly proving The Powerful Influence of Planets and Fixed Stars upon Elementary Bodies, In Justification of the verity of Astrology. Together with an Astrological Judgement Upon The great Conjunction of Saturn and Jupiter* 1603 (London: John Macock, for Nathaniel Brooke, 1650). See note by William Lilly and the prefatory 'To the Reader' by Nicholas Fisk.

80. George Hakewill, *An Apologie or Declaration of the Power and Providence of God in the Government of the World* (3rd ed., Oxford: William Turner, 1635), p. 276.

81. van Helmont, *Oriatrike*, p. 526. On the proposed English society of chemical physicians see Sir Henry Thomas, 'The Society of Chymical Physitians: An Echo of the Great Plague of London—1665', in *Science, Medicine and History*, ed. E. A. Underwood (2 vols., London, 1953), pp. 56–71; P. M. Rattansi 'The Helmontian-Galenist Controversy in Restoration England', *Ambix*, xii (1964), 1–23; C. Webster, 'The English Medical Reformers of the Puritan Revolution: A Background to the 'Society of Chymical Physitians', *Ambix*, xiv (1967), 16–41.

82. George Starkey, *Natures Explication and Helmont's Vindication* (London, 1651); see the 'Epistle Dedicatory'. On Tonstall's challenge, see Robert Wittie's remarks in his *Scarborough Spagirical Anatomiser dissected* (1672). This work is abstracted in Thomas Short, M.D., *The Natural, Experimental, and Medicinal History of the Mineral Waters of Derbyshire, Lincolnshire, and Yorkshire* (London, 1734), pp. 143–8.

Wittie himself was similarly challenged. See F. N. L. Poynter, 'A Seventeenth-Century Medical Controversy: Robert Wittie versus William Simpson', *Science, Medicine and History. Essays on the Evolution of Scientific Thought and Medical Practice written in honour of Charles Singer* (2 vols., London/New York/Toronto: Oxford U.P., 1953). ii, pp. 72–81.

83. John French, *The Art of Distillation* (4th ed., London: T. Williams, 1667), sig A3 recto. From the dedication to Tobias Garband dated London, Nov. 25, 1650.

84. Noah Biggs, Chymiatrophilos, *Mataeotechnia Medicinae Praxews. The Vanity of the Craft of Physicke . . . With an humble Motion for the Reformation of the Universities, And the whole Landscape of Physick, and discovering the Terra Incognita of Chymistrie* (London, 1651), sig. b 1 recto. Biggs was attacked by William Johnson in his preface to Leonard Phioravant's *Three Exact Pieces* (London, 1652), p. 1.

85. J(ohn) H(all), *An Humble Motion To the PARLIAMENT of ENGLAND Concerning The ADVANCEMENT of Learning: and Reformation of the Universities* (London: John Walker, 1649), p. 27.

86. See the article on John Webster by Bertha Potter in the *Dictionary of National Biography*.

87. The work of Hunyades has been discussed by F. Sherwood Taylor and C. H. Josten in 'Johannes Banfi Hunyades 1576–1650', *Ambix*, v (1953), 44–52, and 'Johannes Banfi Hunyades. A Supplementary Note', *Ambix*, v (1956), 115. Here (p. 52) it is suggested that Hunyades arrived in London sometime between 1623 and 1633.

88. John Webster, *Academiarum Examen, or the Examination of Academies. Wherein is discussed and examined the Matter, Method and Customs of Academick and Scholastick Learning, and the insufficiency thereof discovered and laid open; As also some expedients promoting of all kind of Science. Offered to the judgements of all those that love the proficiencie of Arts and Sciences and the advancement of Learning* (London: Giles Calvert, 1654), p. 26.

89. *Ibid.*, p. 71.

90. *Ibid.*, p. 75.

91. *Ibid.*, p. 105.

92. *Ibid.*, pp. 104ff.

93. Seth Ward, *Vindiciae Academiarum containing, Some briefe Animadversions upon Mr Websters Book, Stiled The Examination of Academies. Together with an Appendix concerning what M. Hobbs, and M. Dell have published on this Argument* (Oxford: Leonard Lichfield for Thomas Robinson, 1654), p. 5 From the preface signed N. S. (John Wilkins).

94. *Ibid.*, p. 47.

39

95. *Ibid.*, p. 46.
96. *Ibid.*
97. *Ibid.*
98. Phyllis Allen, *op. cit.*, p. 237.
99. Bacon, *op. cit.*, p. 92.

The research for this paper was completed during the tenure of a Guggenheim Fellowship and with a research grant from the National Institutes of Health (LM 00046).